Cram101 Textbook Outlines to accompany:

Orbital Mechanics for Engineering Students

Howard Curtis, 1st Edition

A Content Technologies Inc. publication (c) 2012.

Cram101 Textbook Outlines and Cram101.com are Cram101 Inc. publications and services. All notes, highlights, reviews, and practice tests are written and prepared by Content Technologies and Cram101, all rights reserved.

STUDYING MADE EASY

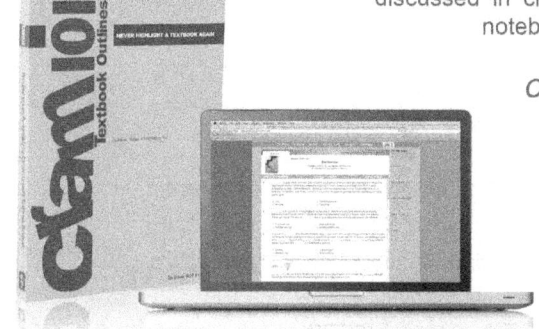

This Cram101 notebook is designed to make studying easier and increase your comprehension of the textbook material. Instead of starting with a blank notebook and trying to write down everything discussed in class lectures, you can use this Cram101 textbook notebook and annotate your notes along with the lecture.

Our goal is to give you the best tools for success.

For a supreme understanding of the course, pair your notebook with our online tools. Should you decide you prefer Cram101.com as your study tool,

we'd like to offer you a trade...

Our Trade In program is a simple way for us to keep our promise and provide you the best studying tools, regardless of where you purchased your Cram101 textbook notebook. As long as your notebook is in *Like New Condition**, you can send it back to us and we will immediately give you a Cram101.com account free for 120 days!

Let The *Trade In* Begin!

THREE SIMPLE STEPS TO TRADE:

1. Go to www.cram101.com/tradein and fill out the packing slip information.
2. Submit and print the packing slip and mail it in with your Cram101 textbook notebook.
3. Activate your account after you receive your email confirmation.

* Books must be returned in *Like New Condition*, meaning there is no damage to the book including, but not limited to; ripped or torn pages, markings or writing on pages, or folded / creased pages. Upon receiving the book, Cram101 will inspect it and reserves the right to terminate your free Cram101.com account and return your textbook notebook at the owners expense.

Learning System

Cram101 Textbook Outlines is a learning system. The notes in this book are the highlights of your textbook, you will never have to highlight a book again.

How to use this book. Take this book to class, it is your notebook for the lecture. The notes and highlights on the left hand side of the pages follow the outline and order of the textbook. All you have to do is follow along while your instructor presents the lecture. Circle the items emphasized in class and add other important information on the right side. With Cram101 Textbook Outlines you'll spend less time writing and more time listening. Learning becomes more efficient.

Cram101.com Online

Increase your studying efficiency by using Cram101.com's practice tests and online reference material. It is the perfect complement to Cram101 Textbook Outlines. Use self-teaching matching tests or simulate in-class testing with comprehensive multiple choice tests, or simply use Cram's true and false tests for quick review. Cram101.com even allows you to enter your in-class notes for an integrated studying format combining the textbook notes with your class notes.

Visit **www.Cram101.com**, click Sign Up at the top of the screen, and enter **DK73DW7255** in the promo code box on the registration screen. Your access to www.Cram101.com is discounted by 50% because you have purchased this book. Sign up and stop highlighting textbooks forever.

Copyright © 2011 by Cram101, Inc. All rights reserved. "Cram101"® and "Never Highlight a Book Again!"® are registered trademarks of Cram101, Inc. ISBN(s): 1428886869. PUBE-4 .2011513

Orbital Mechanics for Engineering Students
Howard Curtis, 1st

CONTENTS

1. DYNAMICS OF POINT MASSES 2
2. tHE TWO-BODY PROBLEM 10
3. ORBITAL POSITION AS A FUNCTION OF TIME 26
4. ORBITS IN THREE DIMENSIONS 34
5. PRELIMINARY ORBIT DETERMINATION 46
6. ORBITAL MANEUVERS 52
7. RELATIVE MOTION AND RENDEZVOUS 64
8. INTERPLANETARY TRAJECTORIES 70
9. RIGID-BODY DYNAMICS 86
10. SATELLITE ATTITUDE DYNAMICS 98
11. ROCKET VEHICLE DYNAMICS 110

Chapter 1. DYNAMICS OF POINT MASSES

Hypothetico-deductive model	The hypothetico-deductive model, first so-named by William Whewell, is a proposed description of scientific method. According to it, scientific inquiry proceeds by formulating a hypothesis in a form that could conceivably be falsified by a test on observable data. A test that could and does run contrary to predictions of the hypothesis is taken as a falsification of the hypothesis.
Motion	In physics, motion is a change in position of an object with respect to time. Change in action is the result of an unapplied force. Motion is typically described in terms of velocity, acceleration, displacement, and time.
Kinematics	It is natural to begin this discussion by considering the various possible types of motion in themselves, leaving out of account for a time the causes to which the initiation of motion may be ascribed; this preliminary enquiry constitutes the science of Kinematics. --ET Whittaker Kinematics is the branch of classical mechanics that describes the motion of bodies (objects) and systems (groups of objects) without consideration of the forces that cause the motion. Kinematics is not to be confused with another branch of classical mechanics: analytical dynamics (the study of the relationship between the motion of objects and its causes), sometimes subdivided into kinetics (the study of the relation between external forces and motion) and statics (the study of the relations in a system at equilibrium). Kinematics also differs from dynamics as used in modern-day physics to describe time-evolution of a system.
Gravitation	In physics, Gravitation is a very important reference book on Einstein's theory of gravity by Charles W. Misner, Kip S. Thorne, and John Archibald Wheeler. Often considered the "Bible" of General Relativity by researchers for its prominence, it is frequently called MTW after its authors' initials, or "the Phone Book" due to its immense size. It was originally published by W. H. Freeman and Company in 1973.

Chapter 1. DYNAMICS OF POINT MASSES

Chapter 1. DYNAMICS OF POINT MASSES

Force	In physics, a force is any influence that causes a free body to undergo a change in speed, a change in direction, or a change in shape. Force can also be described by intuitive concepts such as a push or pull that can cause an object with mass to change its velocity (which includes to begin moving from a state of rest), i.e., to accelerate, or which can cause a flexible object to deform. A force has both magnitude and direction, making it a vector quantity.
Force	In physics, a force is any influence that causes a free body to undergo a change in speed, a change in direction, or a change in shape. Force can also be described by intuitive concepts such as a push or pull that can cause an object with mass to change its velocity (which includes to begin moving from a state of rest), i.e., to accelerate, or which can cause a flexible object to deform. A force has both magnitude and direction, making it a vector quantity.
Impulse	In classical mechanics, an impulse is defined as the integral of a force with respect to time. When a force is applied to a rigid body it changes the momentum of that body. A small force applied for a long time can produce the same momentum change as a large force applied briefly, because it is the product of the force and the time for which it is applied that is important.
Derivative	In calculus, a branch of mathematics, the derivative is a measure of how a function changes as its input changes. Loosely speaking, a derivative can be thought of as how much one quantity is changing in response to changes in some other quantity; for example, the derivative of the position of a moving object with respect to time is the object's instantaneous velocity. Conversely, the integral of the object's velocity over time is how much the object's position changes from the time when the integral begins to the time when the integral ends.
Time derivative	A time derivative is a derivative of a function with respect to time, usually interpreted as the rate of change of the value of the function. The variable denoting time is usually written as t. Notation A variety of notations are used to denote the time derivative.
Leonhard Euler	Leonhard Euler was a pioneering Swiss mathematician and physicist. He made important discoveries in fields as diverse as infinitesimal calculus and graph theory. He also introduced much of the modern mathematical terminology and notation, particularly for mathematical analysis, such as the notion of a mathematical function.

Chapter 1. DYNAMICS OF POINT MASSES

Chapter 1. DYNAMICS OF POINT MASSES

Circular orbit	A circular orbit is the orbit of any point of an object rotating around a fixed axis.
	Below we consider a circular orbit in astrodynamics or celestial mechanics under standard assumptions. Here the centripetal force is the gravitational force, and the axis mentioned above is the line through the center of the central mass perpendicular to the plane of motion.
Linearization	In mathematics and its applications, linearization refers to finding the linear approximation to a function at a given point. In the study of dynamical systems, linearization is a method for assessing the local stability of an equilibrium point of a system of nonlinear differential equations or discrete dynamical systems. This method is used in fields such as engineering, physics, economics, and ecology.
Orbit	In mathematics, in the study of dynamical systems, an orbit is a collection of points related by the evolution function of the dynamical system. The orbit is a subset of the phase space and the set of all orbits is a partition of the phase space, that is different orbits do not intersect in the phase space. Understanding the properties of orbits by using topological method is one of the objectives of the modern theory of dynamical systems.
Relative velocity	In non-relativistic kinematics, relative velocity is the vector difference between the velocities of two objects, as evaluated in terms of a single coordinate system.
	For example, if the velocities of particles A and B are $\mathbf{V}_A$ and $\mathbf{V}_B$ respectively in terms of a given coordinate system, then the relative velocity of A with respect to B (also called the velocity of A relative to B, $\mathbf{V}_{A/B}$, or $\mathbf{V}_{A \text{ rel } B}$) is $$\mathbf{V}_{A \text{ rel } B} = \mathbf{V}_A - \mathbf{V}_B.$$ Conversely, the velocity of B relative to A is $$\mathbf{V}_{B \text{ rel } A} = \mathbf{V}_B - \mathbf{V}_A.$$

Chapter 1. DYNAMICS OF POINT MASSES

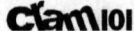

Chapter 1. DYNAMICS OF POINT MASSES

The expression "the velocity of A relative to B" is shorthand for "the velocity of A in the coordinate system where B is always at rest".

Chapter 1. DYNAMICS OF POINT MASSES

Chapter 2. tHE TWO-BODY PROBLEM

Gravitation — In physics, Gravitation is a very important reference book on Einstein's theory of gravity by Charles W. Misner, Kip S. Thorne, and John Archibald Wheeler. Often considered the "Bible" of General Relativity by researchers for its prominence, it is frequently called MTW after its authors' initials, or "the Phone Book" due to its immense size. It was originally published by W. H. Freeman and Company in 1973.

Inertia — Inertia is the resistance of any physical object to a change in its state of motion or rest. It is represented numerically by an object's mass. The principle of inertia is one of the fundamental principles of classical physics which are used to describe the motion of matter and how it is affected by applied forces.

Angular momentum — In physics, angular momentum, moment of momentum, or rotational momentum is a conserved vector quantity that can be used to describe the overall state of a physical system. The angular momentum L of a particle with respect to some point of origin is

$$\mathbf{L} = \mathbf{r} \times \mathbf{p} = \mathbf{r} \times m\mathbf{v},$$

where r is the particle's position from the origin, p = mv is its linear momentum, and × denotes the cross product.

The angular momentum of a system of particles (e.g. a rigid body) is the sum of angular momenta of the individual particles.

Angular velocity — In physics, the angular velocity is a vector quantity (more precisely, a pseudovector) which specifies the angular speed of an object and the axis about which the object is rotating. The SI unit of angular velocity is radians per second, although it may be measured in other units such as degrees per second, revolutions per second, degrees per hour, etc. When measured in cycles or rotations per unit time (e.g. revolutions per minute), it is often called the rotational velocity and its magnitude the rotational speed.

Chapter 2. tHE TWO-BODY PROBLEM

Chapter 2. tHE TWO-BODY PROBLEM

Equations of motion	Equations of motion are equations that describe the behavior of a system (e.g., the motion of a particle under the influence of a force) as a function of time. Sometimes the term refers to the differential equations that the system satisfies (e.g., Newton's second law or Euler-Lagrange equations), and sometimes to the solutions to those equations. Equations of uniformly accelerated linear motion The equations that apply to bodies moving linearly (in one dimension) with constant acceleration are often referred to as "SUVAT" equations where the five variables are represented by those letters (s = displacement, u = initial velocity, v = final velocity, a = acceleration, t = time); the five letters may be shown in a different order.
Hypothetico-deductive model	The hypothetico-deductive model, first so-named by William Whewell, is a proposed description of scientific method. According to it, scientific inquiry proceeds by formulating a hypothesis in a form that could conceivably be falsified by a test on observable data. A test that could and does run contrary to predictions of the hypothesis is taken as a falsification of the hypothesis.
Momentum	In classical mechanics, Momentum is the product of the mass and velocity of an object . In relativistic mechanics, this quantity is multiplied by the Lorentz factor. Momentum is sometimes referred to as linear Momentum to distinguish it from the related subject of angular Momentum.
Motion	In physics, motion is a change in position of an object with respect to time. Change in action is the result of an unapplied force. Motion is typically described in terms of velocity, acceleration, displacement, and time.
Circular orbit	A circular orbit is the orbit of any point of an object rotating around a fixed axis. Below we consider a circular orbit in astrodynamics or celestial mechanics under standard assumptions. Here the centripetal force is the gravitational force, and the axis mentioned above is the line through the center of the central mass perpendicular to the plane of motion.

Chapter 2. tHE TWO-BODY PROBLEM

Chapter 2. tHE TWO-BODY PROBLEM

Orbit	In mathematics, in the study of dynamical systems, an orbit is a collection of points related by the evolution function of the dynamical system. The orbit is a subset of the phase space and the set of all orbits is a partition of the phase space, that is different orbits do not intersect in the phase space. Understanding the properties of orbits by using topological method is one of the objectives of the modern theory of dynamical systems.
Operator	An operator is a mapping from one vector space or module to another. Operators are of critical importance to both linear algebra and functional analysis, and they find application in many other fields of pure and applied mathematics. For example, in classical mechanics the derivative is used ubiquitously, and in quantum mechanics observables are modeled as linear operators.
Linearization	In mathematics and its applications, linearization refers to finding the linear approximation to a function at a given point. In the study of dynamical systems, linearization is a method for assessing the local stability of an equilibrium point of a system of nonlinear differential equations or discrete dynamical systems. This method is used in fields such as engineering, physics, economics, and ecology.
Relative velocity	In non-relativistic kinematics, relative velocity is the vector difference between the velocities of two objects, as evaluated in terms of a single coordinate system.

For example, if the velocities of particles A and B are $\mathbf{V}_A$ and $\mathbf{V}_B$ respectively in terms of a given coordinate system, then the relative velocity of A with respect to B (also called the velocity of A relative to B, $\mathbf{V}_{A/B}$, or $\mathbf{V}_{A \text{ rel } B}$) is

$$\mathbf{V}_{A \text{ rel } B} = \mathbf{V}_A - \mathbf{V}_B.$$

Conversely, the velocity of B relative to A is

$$\mathbf{V}_{B \text{ rel } A} = \mathbf{V}_B - \mathbf{V}_A.$$

The expression "the velocity of A relative to B" is shorthand for "the velocity of A in the coordinate system where B is always at rest".

Chapter 2. tHE TWO-BODY PROBLEM

Chapter 2. tHE TWO-BODY PROBLEM

Eccentricity	In mathematics, the eccentricity, denoted e or ε, is a parameter associated with every conic section. It can be thought of as a measure of how much the conic section deviates from being circular. In particular, - The eccentricity of a circle is zero. - The eccentricity of an ellipse which is not a circle is greater than zero but less than 1. - The eccentricity of a parabola is 1. - The eccentricity of a hyperbola is greater than 1. Furthermore, two conic sections are similar if and only if they have the same eccentricity.
Horizon	There are several types of spacetime horizons that play a role in Einstein's theory of general relativity: - Absolute horizon, a boundary in spacetime in general relativity inside of which events cannot affect an external observer - Apparent horizon, a surface defined in general relativity - Cauchy horizon, a surface found in the study of Cauchy problems - Celestial horizon, a great circle parallel to the horizon - Cosmological horizon, a limit of observability - Event horizon, a boundary in spacetime beyond which events cannot affect the observer - Killing horizon, a null surface on which there is a killing vector field - Particle horizon, the maximum distance from which particles can have travelled to an observer in the age of the universe
Circular orbit	A circular orbit is the orbit of any point of an object rotating around a fixed axis.

Chapter 2. tHE TWO-BODY PROBLEM

Chapter 2. tHE TWO-BODY PROBLEM

Below we consider a circular orbit in astrodynamics or celestial mechanics under standard assumptions. Here the centripetal force is the gravitational force, and the axis mentioned above is the line through the center of the central mass perpendicular to the plane of motion.

| Orbit | In mathematics, in the study of dynamical systems, an orbit is a collection of points related by the evolution function of the dynamical system. The orbit is a subset of the phase space and the set of all orbits is a partition of the phase space, that is different orbits do not intersect in the phase space. Understanding the properties of orbits by using topological method is one of the objectives of the modern theory of dynamical systems. |

| Low Earth orbit | A low Earth orbit is generally defined as an orbit within the locus extending from the Earth's surface up to an altitude of 2,000 km. Given the rapid orbital decay of objects below approximately 200 km, the commonly accepted definition for Low Earth orbit is between 160 - 2,000 km (100 - 1,240 miles) above the Earth's surface. |

With the exception of the lunar flights of the Apollo program, all human spaceflights have either been orbital in Low Earth orbit or sub-orbital.

| Altitude | Altitude is used (aviation, geometry, geographical survey, sport, and more). As a general definition, altitude is a distance measurement, usually in the vertical or "up" direction, between a reference datum and a point or object. The reference datum also often varies according to the context. |

| Flattening | The flattening, ellipticity, or oblateness of an oblate spheroid is a measure of the "squashing" of the spheroid's pole, towards its equator. If a is the distance from the spheroid center to the equator and b the distance from the center to the pole then |

$$flattening = \frac{a-b}{a}$$

First, second and third flattening

Chapter 2. tHE TWO-BODY PROBLEM

Chapter 2. tHE TWO-BODY PROBLEM

The first, primary flattening, f, is the versine of the spheroid's angular eccentricity (" α "), equalling the relative difference between its equatorial radius, a, and its polar radius, b :

$$f = \operatorname{ver}(\alpha) = 2\sin^2\left(\frac{\alpha}{2}\right) = 1 - \cos(\alpha) = \frac{a-b}{a};$$

- The flattening of the Earth in WGS-84 is 1:298.257223563 (which corresponds to a radius difference of 21.385 km (13 mi) of the Earth radius 6378.137 - 6356.752 km) and would not be realized visually from space, since the difference represents only 0.335 %.
- The flattening of Jupiter (1:16) and Saturn (nearly 1:10), in contrast, can be seen even in a small telescope;
- Conversely, that of the Sun is less than 1:1000 and that of the Moon barely 1:900.

The amount of flattening depends on

and in detail on

- size and density of the celestial body ;
- the rotation of the planet or star;
- and the elasticity of the body.

There is also a second flattening, f',

$$f' = \frac{2\sin^2(\alpha/2)}{1 - 2\sin^2(\alpha/2)} = \frac{a-b}{b}$$

and a third flattening, f '' (sometimes denoted as "n", firstly used the notation in 1837 by Friedrich Bessel on calculation of meridian arc length), that is the squared half-angle tangent of α :

$$f'' = n = \tan^2\left(\frac{\alpha}{2}\right) = \frac{1 - \cos(\alpha)}{1 + \cos(\alpha)} = \frac{a-b}{a+b};$$

Chapter 2. tHE TWO-BODY PROBLEM

Chapter 2. tHE TWO-BODY PROBLEM

Escape velocity

In physics, escape velocity is the speed at which the kinetic energy plus the gravitational potential energy of an object is zero. It is the speed needed to "break free" from a gravitational field without further propulsion. The term escape velocity is actually a misnomer, as the concept refers to a scalar speed which is independent of direction.

Delta-v

In astrodynamics, Δv or delta-v is a scalar which takes units of speed that measures the amount of "effort" needed to carry out an orbital maneuver, i.e., to change from one trajectory to another.

$$\Delta v = \int_{t_0}^{t_1} \frac{|T|}{m} dt$$

where

    T is the instantaneous thrust
    m is the instantaneous mass

If there are no other external forces than gravity, this is the integral of the magnitude of the g-force.

In the absence of external forces, and when thrust is applied in a constant direction this simplifies to:

$$= \int_{t_0}^{t_1} |a|\, dt = |v_1 - v_0|$$

which is simply the magnitude of the change in velocity.

Orbit equation

In astrodynamics an orbit equation defines the path of orbiting body m_2 around central body m_1 relative to m_1, without specifying position as a function of time. Under standard assumptions, a body moving under the influence of a force, directed to a central body, with a magnitude inversely proportional to the square of the distance (such as gravity), has an orbit that is a conic section with the central body located at one of the two foci, or the focus (Kepler's first law).

Chapter 2. tHE TWO-BODY PROBLEM

Chapter 2. tHE TWO-BODY PROBLEM

If the conic section intersects the central body, then the actual trajectory can only be the part above the surface, but for that part the orbit equation and many related formulas still apply, as long as it is a freefall (situation of weightlessness).

Parabola — In mathematics, the parabola is a conic section, the intersection of a right circular conical surface and a plane parallel to a generating straight line of that surface. Given a point (the focus) and a corresponding line (the directrix) on the plane, the locus of points in that plane that are equidistant from them is a parabola. The line perpendicular to the directrix and passing through the focus is called the "axis of symmetry".

Hyperbola — In mathematics a hyperbola is a curve, specifically a smooth curve that lies in a plane, which can be defined either by its geometric properties or by the kinds of equations for which it is the solution set. A hyperbola has two pieces, called connected components or branches, which are mirror images of each other and resembling two infinite bows. The hyperbola is one of the four kinds of conic section, formed by the intersection of a plane and a cone.

Derivative — In calculus, a branch of mathematics, the derivative is a measure of how a function changes as its input changes. Loosely speaking, a derivative can be thought of as how much one quantity is changing in response to changes in some other quantity; for example, the derivative of the position of a moving object with respect to time is the object's instantaneous velocity. Conversely, the integral of the object's velocity over time is how much the object's position changes from the time when the integral begins to the time when the integral ends.

Chapter 2. tHE TWO-BODY PROBLEM

Chapter 3. ORBITAL POSITION AS A FUNCTION OF TIME

Circular orbit	A circular orbit is the orbit of any point of an object rotating around a fixed axis.
	Below we consider a circular orbit in astrodynamics or celestial mechanics under standard assumptions. Here the centripetal force is the gravitational force, and the axis mentioned above is the line through the center of the central mass perpendicular to the plane of motion.
Orbit	In mathematics, in the study of dynamical systems, an orbit is a collection of points related by the evolution function of the dynamical system. The orbit is a subset of the phase space and the set of all orbits is a partition of the phase space, that is different orbits do not intersect in the phase space. Understanding the properties of orbits by using topological method is one of the objectives of the modern theory of dynamical systems.
Motion	In physics, motion is a change in position of an object with respect to time. Change in action is the result of an unapplied force. Motion is typically described in terms of velocity, acceleration, displacement, and time.
Eccentricity	In mathematics, the eccentricity, denoted e or ε, is a parameter associated with every conic section. It can be thought of as a measure of how much the conic section deviates from being circular.
	In particular,
	- The eccentricity of a circle is zero.
- The eccentricity of an ellipse which is not a circle is greater than zero but less than 1.
- The eccentricity of a parabola is 1.
- The eccentricity of a hyperbola is greater than 1.
	Furthermore, two conic sections are similar if and only if they have the same eccentricity.
Eccentric anomaly	In celestial mechanics, the eccentric anomaly is an angular parameter that defines the position of a body that is moving along an elliptic Kepler orbit.

Chapter 3. ORBITAL POSITION AS A FUNCTION OF TIME

Chapter 3. ORBITAL POSITION AS A FUNCTION OF TIME

For the point P orbiting around an ellipse, the eccentric anomaly is the angle E in the figure. It is determined by drawing a vertical line from the major axis of the ellipse through the point P and locating its intercept P' with the auxiliary circle, a circle of radius a (the semi-major axis of the ellipse) that enscribes the entire ellipse.

Hypothetico-deductive model

The hypothetico-deductive model, first so-named by William Whewell, is a proposed description of scientific method. According to it, scientific inquiry proceeds by formulating a hypothesis in a form that could conceivably be falsified by a test on observable data. A test that could and does run contrary to predictions of the hypothesis is taken as a falsification of the hypothesis.

Hyperbola

In mathematics a hyperbola is a curve, specifically a smooth curve that lies in a plane, which can be defined either by its geometric properties or by the kinds of equations for which it is the solution set. A hyperbola has two pieces, called connected components or branches, which are mirror images of each other and resembling two infinite bows. The hyperbola is one of the four kinds of conic section, formed by the intersection of a plane and a cone.

Laplace limit

In mathematics, the Laplace limit is the maximum value of the eccentricity for which the series solution to Kepler's equation converges. It is approximately

0.66274 34193 49181 58097 47420 97109 25290.

Kepler's equation M = E − ε sin E relates the mean anomaly M with the eccentric anomaly E for a body moving in an ellipse with eccentricity ε. This equation cannot be solved for E in terms of elementary functions, but the Lagrange reversion theorem yields the solution as a power series in ε:

$$E = M + \sin(M)\varepsilon + \tfrac{1}{2}\sin(2M)\varepsilon^2 + \left(\tfrac{3}{8}\sin(3M) - \tfrac{1}{8}\sin(M)\right)\varepsilon^3 + \cdot$$

Laplace realized that this series converges for small values of the eccentricity, but diverges when the eccentricity exceeds a certain value.

Chapter 3. ORBITAL POSITION AS A FUNCTION OF TIME

Chapter 3. ORBITAL POSITION AS A FUNCTION OF TIME

Hyperbola	In mathematics a hyperbola is a curve, specifically a smooth curve that lies in a plane, which can be defined either by its geometric properties or by the kinds of equations for which it is the solution set. A hyperbola has two pieces, called connected components or branches, which are mirror images of each other and resembling two infinite bows. The hyperbola is one of the four kinds of conic section, formed by the intersection of a plane and a cone.
Orbit determination	Orbit determination is a branch of astronomy specialised in calculating, and hence predicting, the orbits of objects such as moons, planets, and spacecraft . These orbits could be orbiting the Earth, or other bodies. The determination of the orbit of newly observed asteroids is a common usage of these techniques, both so the asteroid can be followed up with future observations, and also to check that it has not been previously discovered.
Stumpff function	In celestial mechanics, the Stumpff functions $c_k(x)$, developed by Karl Stumpff, are used for analyzing orbits using the universal variable formulation. They are defined by the formula: $$c_k(x) = \frac{1}{k!} - \frac{x}{(k+2)!} + \frac{x^2}{(k+4)!} - \ldots = \sum_{i=0}^{\infty} \frac{(-1)^i x^i}{(k+2i)!}$$ for k = 0,1,2,3.... The series above converges absolutely for all real x. By comparing the Taylor series expansion of the trigonometric functions sin and cos with $c_0(x)$ and $c_1(x)$, a relationship can be found: Similarly, by comparing with the expansion of the hyperbolic functions sinh and cosh we find: The Stumpff functions satisfy the recursive relations:
Orbit equation	In astrodynamics an orbit equation defines the path of orbiting body m_2 around central body m_1 relative to m_1, without specifying position as a function of time. Under standard assumptions, a body moving under the influence of a force, directed to a central body, with a magnitude inversely proportional to the square of the distance (such as gravity), has an orbit that is a conic section with the central body located at one of the two foci, or the focus (Kepler's first law).

Chapter 3. ORBITAL POSITION AS A FUNCTION OF TIME

Chapter 3. ORBITAL POSITION AS A FUNCTION OF TIME

If the conic section intersects the central body, then the actual trajectory can only be the part above the surface, but for that part the orbit equation and many related formulas still apply, as long as it is a freefall (situation of weightlessness).

Chapter 3. ORBITAL POSITION AS A FUNCTION OF TIME

Chapter 4. ORBITS IN THREE DIMENSIONS

Orbit	In mathematics, in the study of dynamical systems, an orbit is a collection of points related by the evolution function of the dynamical system. The orbit is a subset of the phase space and the set of all orbits is a partition of the phase space, that is different orbits do not intersect in the phase space. Understanding the properties of orbits by using topological method is one of the objectives of the modern theory of dynamical systems.
Orbit determination	Orbit determination is a branch of astronomy specialised in calculating, and hence predicting, the orbits of objects such as moons, planets, and spacecraft . These orbits could be orbiting the Earth, or other bodies. The determination of the orbit of newly observed asteroids is a common usage of these techniques, both so the asteroid can be followed up with future observations, and also to check that it has not been previously discovered.
Orbit	In mathematics, in the study of dynamical systems, an orbit is a collection of points related by the evolution function of the dynamical system. The orbit is a subset of the phase space and the set of all orbits is a partition of the phase space, that is different orbits do not intersect in the phase space. Understanding the properties of orbits by using topological method is one of the objectives of the modern theory of dynamical systems.
Sphere	A sphere is a perfectly round geometrical object in three-dimensional space, such as the shape of a round ball. Like a circle in two dimensions, a perfect sphere is completely symmetrical around its center, with all points on the surface lying the same distance r from the center point. This distance r is known as the radius of the sphere.
Rotation	In geometry and linear algebra, a rotation is a transformation in a plane or in space that describes the motion of a rigid body around a fixed point. A rotation is different from a translation, which has no fixed points, and from a reflection, which "flips" the bodies it is transforming. A rotation and the above-mentioned transformations are isometries; they leave the distance between any two points unchanged after the transformation.
Equinox	An equinox occurs twice a year, when the tilt of the Earth's axis is inclined neither away from nor towards the Sun, the center of the Sun being in the same plane as the Earth's equator. The term equinox can also be used in a broader sense, meaning the date when such a passage happens. The name "equinox" is derived from the Latin aequus (equal) and nox (night), because around the equinox, the night and day are approximately equally long.
Gravitation	In physics, Gravitation is a very important reference book on Einstein's theory of gravity by Charles W. Misner, Kip S. Thorne, and John Archibald Wheeler. Often considered the "Bible" of General Relativity by researchers for its prominence, it is frequently called MTW after its authors' initials, or "the Phone Book" due to its immense size. It was originally published by W. H. Freeman and Company in 1973.

Chapter 4. ORBITS IN THREE DIMENSIONS

Chapter 4. ORBITS IN THREE DIMENSIONS

Perturbation Perturbation is a term used in astronomy in connection with descriptions of the complex motion of a massive body which is subject to appreciable gravitational effects from more than one other massive body.

Such complex motions of a body can be broken down descriptively into component parts. First, there can be the hypothetical motion that the body would follow, if it moved under the gravitational effect of one other body only.

Ephemeris An ephemeris is a table of values that gives the positions of astronomical objects in the sky at a given time or times. Different kinds of ephemerides are used for astronomy and astrology. Even though this was also one of the first applications of mechanical computers, an ephemeris will still often be a simple printed table.

State In functional analysis, a state on a C*-algebra is a positive linear functional of norm 1. The set of states of a C*-algebra A, sometimes denoted by S(A), is always a convex set. The extremal points of S(A) are called pure states. If A has a multiplicative identity, S(A) is compact in the weak*-topology.

Delta-v In astrodynamics, Δv or delta-v is a scalar which takes units of speed that measures the amount of "effort" needed to carry out an orbital maneuver, i.e., to change from one trajectory to another.

$$\Delta v = \int_{t_0}^{t_1} \frac{|T|}{m}\, dt$$

where

> T is the instantaneous thrust
> m is the instantaneous mass

If there are no other external forces than gravity, this is the integral of the magnitude of the g-force.

Chapter 4. ORBITS IN THREE DIMENSIONS

Chapter 4. ORBITS IN THREE DIMENSIONS

In the absence of external forces, and when thrust is applied in a constant direction this simplifies to:

$$= \int_{t_0}^{t_1} |a|\, dt = |v_1 - v_0|$$

which is simply the magnitude of the change in velocity.

Geocentric orbit

A geocentric orbit involves any object orbiting the Earth, such as the Moon or artificial satellites. Currently there are approximately 2,465 artificial satellites orbiting the Earth and 6,216 pieces of space debris as tracked by the Goddard Space Flight Center. Over 16,291 previously launched objects have decayed into the Earth's atmosphere.

Eccentricity

In mathematics, the eccentricity, denoted e or ε, is a parameter associated with every conic section. It can be thought of as a measure of how much the conic section deviates from being circular.

In particular,

- The eccentricity of a circle is zero.
- The eccentricity of an ellipse which is not a circle is greater than zero but less than 1.
- The eccentricity of a parabola is 1.
- The eccentricity of a hyperbola is greater than 1.

Furthermore, two conic sections are similar if and only if they have the same eccentricity.

Orbital elements

Orbital elements are the parameters required to uniquely identify a specific orbit. In celestial mechanics these elements are generally considered in classical two-body systems, where a Kepler orbit is used (derived from Newton's laws of motion and Newton's law of universal gravitation). There are many different ways to mathematically describe the same orbit, but certain schemes each consisting of a set of six parameters are commonly used in astronomy and orbital mechanics.

Chapter 4. ORBITS IN THREE DIMENSIONS

Chapter 4. ORBITS IN THREE DIMENSIONS

Circular orbit	A circular orbit is the orbit of any point of an object rotating around a fixed axis.
	Below we consider a circular orbit in astrodynamics or celestial mechanics under standard assumptions. Here the centripetal force is the gravitational force, and the axis mentioned above is the line through the center of the central mass perpendicular to the plane of motion.
Sun-synchronous orbit	A sun-synchronous orbit is a geocentric orbit which combines altitude and inclination in such a way that an object on that orbit ascends or descends over any given point of the Earth's surface at the same local mean solar time. The surface illumination angle will be nearly the same every time. This consistent lighting is a useful characteristic for satellites that image the Earth's surface in visible or infrared wavelengths (e.g. weather and spy satellites) and for other remote sensing satellites (e.g. those carrying ocean and atmospheric remote sensing instruments that require sunlight).
Argument	In mathematics, arg is a function operating on complex numbers (visualised as a flat plane). It gives the angle between the line joining the point to the origin and the positive real axis, shown as φ in figure 1 opposite, known as an argument of the point (that is, the angle between the half-lines of the position vector representing the number and the positive real axis).
	Arguments are defined in two equivalent ways:
	• Geometrically, in relation to an Argand diagram, arg z is the angle φ from the positive real axis to the vector representing z.
Hypothetico-deductive model	The hypothetico-deductive model, first so-named by William Whewell, is a proposed description of scientific method. According to it, scientific inquiry proceeds by formulating a hypothesis in a form that could conceivably be falsified by a test on observable data. A test that could and does run contrary to predictions of the hypothesis is taken as a falsification of the hypothesis.

Chapter 4. ORBITS IN THREE DIMENSIONS

Chapter 4. ORBITS IN THREE DIMENSIONS

Orthonormal basis	In mathematics, particularly linear algebra, an orthonormal basis for inner product space V with finite dimension is a basis for V whose vectors are orthonormal. For example, the standard basis for a Euclidean space R^n is an orthonormal basis, where the relevant inner product is the dot product of vectors. The image of the standard basis under a rotation or reflection (or any orthogonal transformation) is also orthonormal, and every orthonormal basis for R^n arises in this fashion.
Stumpff function	In celestial mechanics, the Stumpff functions $c_k(x)$, developed by Karl Stumpff, are used for analyzing orbits using the universal variable formulation. They are defined by the formula: $$c_k(x) = \frac{1}{k!} - \frac{x}{(k+2)!} + \frac{x^2}{(k+4)!} - \ldots = \sum_{i=0}^{\infty} \frac{(-1)^i x^i}{(k+2i)!}$$ for k = 0,1,2,3.... The series above converges absolutely for all real x.
	By comparing the Taylor series expansion of the trigonometric functions sin and cos with $c_0(x)$ and $c_1(x)$, a relationship can be found:
	Similarly, by comparing with the expansion of the hyperbolic functions sinh and cosh we find:
	The Stumpff functions satisfy the recursive relations:
Molniya orbit	A Molniya orbit is a type of highly elliptical orbit with an inclination of 63.4 degrees and an orbital period of precisely one half of a sidereal day. Molniya orbits are named after a series of Soviet/Russian Molniya communications satellites which have been using this type of orbit since the mid 1960s.
	A satellite placed in a Molniya orbit spends most of its time over a designated area of the earth as a result of "apogee dwell".

Chapter 4. ORBITS IN THREE DIMENSIONS

Chapter 4. ORBITS IN THREE DIMENSIONS

Motion	In physics, motion is a change in position of an object with respect to time. Change in action is the result of an unapplied force. Motion is typically described in terms of velocity, acceleration, displacement, and time.

Chapter 4. ORBITS IN THREE DIMENSIONS

Chapter 5. PRELIMINARY ORBIT DETERMINATION

Orbit	In mathematics, in the study of dynamical systems, an orbit is a collection of points related by the evolution function of the dynamical system. The orbit is a subset of the phase space and the set of all orbits is a partition of the phase space, that is different orbits do not intersect in the phase space. Understanding the properties of orbits by using topological method is one of the objectives of the modern theory of dynamical systems.
Orbit determination	Orbit determination is a branch of astronomy specialised in calculating, and hence predicting, the orbits of objects such as moons, planets, and spacecraft . These orbits could be orbiting the Earth, or other bodies. The determination of the orbit of newly observed asteroids is a common usage of these techniques, both so the asteroid can be followed up with future observations, and also to check that it has not been previously discovered.
Hypothetico-deductive model	The hypothetico-deductive model, first so-named by William Whewell, is a proposed description of scientific method. According to it, scientific inquiry proceeds by formulating a hypothesis in a form that could conceivably be falsified by a test on observable data. A test that could and does run contrary to predictions of the hypothesis is taken as a falsification of the hypothesis.
Orbital elements	Orbital elements are the parameters required to uniquely identify a specific orbit. In celestial mechanics these elements are generally considered in classical two-body systems, where a Kepler orbit is used (derived from Newton's laws of motion and Newton's law of universal gravitation). There are many different ways to mathematically describe the same orbit, but certain schemes each consisting of a set of six parameters are commonly used in astronomy and orbital mechanics.
Precession	Precession is the process of a round part in a round hole rotating with respect to that hole because of clearance between between them and a radial force on the part than changes direction. The direction of rotation of the inner part is opposite to the direction of rotation of the radial force. Fretting between the part and the hole is often a result of this motion.
Orbit	In mathematics, in the study of dynamical systems, an orbit is a collection of points related by the evolution function of the dynamical system. The orbit is a subset of the phase space and the set of all orbits is a partition of the phase space, that is different orbits do not intersect in the phase space. Understanding the properties of orbits by using topological method is one of the objectives of the modern theory of dynamical systems.

Chapter 5. PRELIMINARY ORBIT DETERMINATION

Chapter 5. PRELIMINARY ORBIT DETERMINATION

Stumpff function

In celestial mechanics, the Stumpff functions $c_k(x)$, developed by Karl Stumpff, are used for analyzing orbits using the universal variable formulation. They are defined by the formula:

$$c_k(x) = \frac{1}{k!} - \frac{x}{(k+2)!} + \frac{x^2}{(k+4)!} - \cdots = \sum_{i=0}^{\infty} \frac{(-1)^i x^i}{(k+2i)!}$$

for k = 0,1,2,3.... The series above converges absolutely for all real x.

By comparing the Taylor series expansion of the trigonometric functions sin and cos with $c_0(x)$ and $c_1(x)$, a relationship can be found:

Similarly, by comparing with the expansion of the hyperbolic functions sinh and cosh we find:

The Stumpff functions satisfy the recursive relations:

Altitude

Altitude is used (aviation, geometry, geographical survey, sport, and more). As a general definition, altitude is a distance measurement, usually in the vertical or "up" direction, between a reference datum and a point or object. The reference datum also often varies according to the context.

Eccentricity

In mathematics, the eccentricity, denoted e or ε, is a parameter associated with every conic section. It can be thought of as a measure of how much the conic section deviates from being circular.

In particular,

- The eccentricity of a circle is zero.
- The eccentricity of an ellipse which is not a circle is greater than zero but less than 1.
- The eccentricity of a parabola is 1.
- The eccentricity of a hyperbola is greater than 1.

Chapter 5. PRELIMINARY ORBIT DETERMINATION

Chapter 5. PRELIMINARY ORBIT DETERMINATION

	Furthermore, two conic sections are similar if and only if they have the same eccentricity.
Horizon	There are several types of spacetime horizons that play a role in Einstein's theory of general relativity: • Absolute horizon, a boundary in spacetime in general relativity inside of which events cannot affect an external observer • Apparent horizon, a surface defined in general relativity • Cauchy horizon, a surface found in the study of Cauchy problems • Celestial horizon, a great circle parallel to the horizon • Cosmological horizon, a limit of observability • Event horizon, a boundary in spacetime beyond which events cannot affect the observer • Killing horizon, a null surface on which there is a killing vector field • Particle horizon, the maximum distance from which particles can have travelled to an observer in the age of the universe
State	In functional analysis, a state on a C*-algebra is a positive linear functional of norm 1. The set of states of a C*-algebra A, sometimes denoted by S(A), is always a convex set. The extremal points of S(A) are called pure states. If A has a multiplicative identity, S(A) is compact in the weak*-topology.
Motion	In physics, motion is a change in position of an object with respect to time. Change in action is the result of an unapplied force. Motion is typically described in terms of velocity, acceleration, displacement, and time.

Chapter 5. PRELIMINARY ORBIT DETERMINATION

Chapter 6. ORBITAL MANEUVERS

Orbital maneuver | In spaceflight, an orbital maneuver is the use of propulsion systems to change the orbit of a spacecraft. For spacecraft far from Earth--for example those in orbits around the Sun--an orbital maneuver is called a deep-space maneuver (DSM).

Impulsive maneuvers

An "impulsive maneuver" is one which involves a single, ideally instantaneous change in the spacecraft's velocity.

Lagrange multiplier | In mathematical optimization, the method of Lagrange multipliers provides a strategy for finding the maxima and minima of a function subject to constraints.

For instance, consider the optimization problem

$$\text{maximize } f(x,y)$$
$$\text{subject to } g(x,y) = c.$$

We introduce a new variable (λ) called a Lagrange multiplier, and study the Lagrange function defined by

$$\Lambda(x,y,\lambda) = f(x,y) + \lambda \cdot \big(g(x,y) - c\big).$$

(the λ term may be either added or subtracted). If (x,y) is a maximum for the original constrained problem, then there exists a λ such that (x,y,λ) is a stationary point for the Lagrange function (stationary points are those points where the partial derivatives of Λ are zero).

Chapter 6. ORBITAL MANEUVERS

Chapter 6. ORBITAL MANEUVERS

Specific impulse	Specific impulse is a way to describe the efficiency of rocket and jet engines. It represents the impulse (change in momentum) per unit amount of propellant used. The unit amount may be given either per unit mass (such as kilograms), or per unit Earth-weight (such as kiloponds, since g is used for the latter definition).
Impulse	In classical mechanics, an impulse is defined as the integral of a force with respect to time. When a force is applied to a rigid body it changes the momentum of that body. A small force applied for a long time can produce the same momentum change as a large force applied briefly, because it is the product of the force and the time for which it is applied that is important.
Hypothetico-deductive model	The hypothetico-deductive model, first so-named by William Whewell, is a proposed description of scientific method. According to it, scientific inquiry proceeds by formulating a hypothesis in a form that could conceivably be falsified by a test on observable data. A test that could and does run contrary to predictions of the hypothesis is taken as a falsification of the hypothesis.
Multiplier	In Fourier analysis, a multiplier operator is a type of linear operator, or transformation of functions. These operators act on a function by altering its Fourier transform. Specifically they multiply the Fourier transform of a function by a specified function known as the multiplier or symbol.
Bi-elliptic transfer	In astronautics and aerospace engineering, the bi-elliptic transfer is an orbital maneuver that moves a spacecraft from one orbit to another and may, in certain situations, require less delta-v than a Hohmann transfer. The bi-elliptic transfer consists of two half elliptic orbits. From the initial orbit, a delta-v is applied boosting the spacecraft into the first transfer orbit with an apoapsis at some point r_b away from the central body.
Orbit	In mathematics, in the study of dynamical systems, an orbit is a collection of points related by the evolution function of the dynamical system. The orbit is a subset of the phase space and the set of all orbits is a partition of the phase space, that is different orbits do not intersect in the phase space. Understanding the properties of orbits by using topological method is one of the objectives of the modern theory of dynamical systems.

Chapter 6. ORBITAL MANEUVERS

Chapter 6. ORBITAL MANEUVERS

Elliptic orbit

In astrodynamics or celestial mechanics an elliptic orbit is a Kepler orbit with the eccentricity less than 1; this includes the special case of a circular orbit, with eccentricity equal to zero. In a stricter sense, it is a Kepler orbit with the eccentricity greater than 0 and less than 1 (thus excluding the circular orbit). In a wider sense it is a Kepler orbit with negative energy.

Delta-v

In astrodynamics, Δv or delta-v is a scalar which takes units of speed that measures the amount of "effort" needed to carry out an orbital maneuver, i.e., to change from one trajectory to another.

$$\Delta v = \int_{t_0}^{t_1} \frac{|T|}{m} \, dt$$

where

T is the instantaneous thrust
m is the instantaneous mass

If there are no other external forces than gravity, this is the integral of the magnitude of the g-force.

In the absence of external forces, and when thrust is applied in a constant direction this simplifies to:

$$= \int_{t_0}^{t_1} |a| \, dt = |v_1 - v_0|$$

which is simply the magnitude of the change in velocity.

Gravitation

In physics, Gravitation is a very important reference book on Einstein's theory of gravity by Charles W. Misner, Kip S. Thorne, and John Archibald Wheeler. Often considered the "Bible" of General Relativity by researchers for its prominence, it is frequently called MTW after its authors' initials, or "the Phone Book" due to its immense size. It was originally published by W. H. Freeman and Company in 1973.

Chapter 6. ORBITAL MANEUVERS

Chapter 6. ORBITAL MANEUVERS

Rotation	In geometry and linear algebra, a rotation is a transformation in a plane or in space that describes the motion of a rigid body around a fixed point. A rotation is different from a translation, which has no fixed points, and from a reflection, which "flips" the bodies it is transforming. A rotation and the above-mentioned transformations are isometries; they leave the distance between any two points unchanged after the transformation.
Eccentricity	In mathematics, the eccentricity, denoted e or ε, is a parameter associated with every conic section. It can be thought of as a measure of how much the conic section deviates from being circular. In particular, - The eccentricity of a circle is zero. - The eccentricity of an ellipse which is not a circle is greater than zero but less than 1. - The eccentricity of a parabola is 1. - The eccentricity of a hyperbola is greater than 1. Furthermore, two conic sections are similar if and only if they have the same eccentricity.
True anomaly	In celestial mechanics, the true anomaly is an angular parameter that defines the position of a body moving along a Keplerian orbit. It is the angle between the direction of periapsis and the current position of the body, as seen from the main focus of the ellipse (the point around which the object orbits). The true anomaly is usually denoted by the Greek letters ν or θ, or the Roman letter f.
Angular momentum	In physics, angular momentum, moment of momentum, or rotational momentum is a conserved vector quantity that can be used to describe the overall state of a physical system. The angular momentum L of a particle with respect to some point of origin is $$\mathbf{L} = \mathbf{r} \times \mathbf{p} = \mathbf{r} \times m\mathbf{v},$$

Chapter 6. ORBITAL MANEUVERS

Chapter 6. ORBITAL MANEUVERS

where r is the particle's position from the origin, p = mv is its linear momentum, and × denotes the cross product.

The angular momentum of a system of particles (e.g. a rigid body) is the sum of angular momenta of the individual particles.

Momentum | In classical mechanics, Momentum is the product of the mass and velocity of an object. In relativistic mechanics, this quantity is multiplied by the Lorentz factor. Momentum is sometimes referred to as linear Momentum to distinguish it from the related subject of angular Momentum.

Rotation | In geometry and linear algebra, a rotation is a transformation in a plane or in space that describes the motion of a rigid body around a fixed point. A rotation is different from a translation, which has no fixed points, and from a reflection, which "flips" the bodies it is transforming. A rotation and the above-mentioned transformations are isometries; they leave the distance between any two points unchanged after the transformation.

Inclination | Inclination in general is the angle between a reference plane and another plane or axis of direction.

Orbits

The inclination is one of the six orbital parameters describing the shape and orientation of a celestial orbit. It is the angular distance of the orbital plane from the plane of reference (usually the primary's equator or the ecliptic), normally stated in degrees.

Ground track | A ground track is the path on the surface of the Earth directly below an aircraft or satellite. In the case of a satellite, it is the projection of the satellite's orbit onto the surface of the Earth (or whatever body the satellite is orbiting).

Chapter 6. ORBITAL MANEUVERS

Chapter 6. ORBITAL MANEUVERS

	A satellite ground track may be thought of as a path along the Earth's surface which traces the movement of an imaginary line between the satellite and the center of the Earth.
Projection	In linear algebra and functional analysis, a projection is a linear transformation P from a vector space to itself such that $P^2 = P$. It leaves its image unchanged. Though abstract, this definition of "projection" formalizes and generalizes the idea of graphical projection. One can also consider the effect of a projection on a geometrical object by examining the effect of the projection on points in the object.
Low Earth orbit	A low Earth orbit is generally defined as an orbit within the locus extending from the Earth's surface up to an altitude of 2,000 km. Given the rapid orbital decay of objects below approximately 200 km, the commonly accepted definition for Low Earth orbit is between 160 - 2,000 km (100 - 1,240 miles) above the Earth's surface. With the exception of the lunar flights of the Apollo program, all human spaceflights have either been orbital in Low Earth orbit or sub-orbital.
Ellipse	In geometry, an ellipse is a plane curve that results from the intersection of a cone by a plane in a way that produces a closed curve. Circles are special cases of ellipses, obtained when the cutting plane is orthogonal to the cone's axis. An ellipse is also the locus of all points of the plane whose distances to two fixed points add to the same constant.

Chapter 6. ORBITAL MANEUVERS

Chapter 7. RELATIVE MOTION AND RENDEZVOUS

Motion	In physics, motion is a change in position of an object with respect to time. Change in action is the result of an unapplied force. Motion is typically described in terms of velocity, acceleration, displacement, and time.
Relative velocity	In non-relativistic kinematics, relative velocity is the vector difference between the velocities of two objects, as evaluated in terms of a single coordinate system.

For example, if the velocities of particles A and B are $\mathbf{V}_A$ and $\mathbf{V}_B$ respectively in terms of a given coordinate system, then the relative velocity of A with respect to B (also called the velocity of A relative to B, $\mathbf{V}_{A/B}$, or $\mathbf{V}_A$ rel B) is

$$\mathbf{V}_{A \text{ rel } B} = \mathbf{V}_A - \mathbf{V}_B.$$

Conversely, the velocity of B relative to A is

$$\mathbf{V}_{B \text{ rel } A} = \mathbf{V}_B - \mathbf{V}_A.$$

The expression "the velocity of A relative to B" is shorthand for "the velocity of A in the coordinate system where B is always at rest".

Linearization	In mathematics and its applications, linearization refers to finding the linear approximation to a function at a given point. In the study of dynamical systems, linearization is a method for assessing the local stability of an equilibrium point of a system of nonlinear differential equations or discrete dynamical systems. This method is used in fields such as engineering, physics, economics, and ecology.
Angular momentum	In physics, angular momentum, moment of momentum, or rotational momentum is a conserved vector quantity that can be used to describe the overall state of a physical system. The angular momentum L of a particle with respect to some point of origin is

$$\mathbf{L} = \mathbf{r} \times \mathbf{p} = \mathbf{r} \times m\mathbf{v},$$

Chapter 7. RELATIVE MOTION AND RENDEZVOUS

Chapter 7. RELATIVE MOTION AND RENDEZVOUS

where r is the particle's position from the origin, p = mv is its linear momentum, and × denotes the cross product.

The angular momentum of a system of particles (e.g. a rigid body) is the sum of angular momenta of the individual particles.

Momentum	In classical mechanics, Momentum is the product of the mass and velocity of an object. In relativistic mechanics, this quantity is multiplied by the Lorentz factor. Momentum is sometimes referred to as linear Momentum to distinguish it from the related subject of angular Momentum.
Differential	In calculus, a differential is traditionally an infinitesimally small change in a variable. For example, if x is a variable, then a change in the value of x is often denoted Δx (or δx when this change is considered to be small). The differential dx represents such a change, but is infinitely small.
Impulse	In classical mechanics, an impulse is defined as the integral of a force with respect to time. When a force is applied to a rigid body it changes the momentum of that body. A small force applied for a long time can produce the same momentum change as a large force applied briefly, because it is the product of the force and the time for which it is applied that is important.
Orbital maneuver	In spaceflight, an orbital maneuver is the use of propulsion systems to change the orbit of a spacecraft. For spacecraft far from Earth--for example those in orbits around the Sun--an orbital maneuver is called a deep-space maneuver (DSM). Impulsive maneuvers An "impulsive maneuver" is one which involves a single, ideally instantaneous change in the spacecraft's velocity.
Circular orbit	A circular orbit is the orbit of any point of an object rotating around a fixed axis.

Chapter 7. RELATIVE MOTION AND RENDEZVOUS

Chapter 7. RELATIVE MOTION AND RENDEZVOUS

	Below we consider a circular orbit in astrodynamics or celestial mechanics under standard assumptions. Here the centripetal force is the gravitational force, and the axis mentioned above is the line through the center of the central mass perpendicular to the plane of motion.
Orbit	In mathematics, in the study of dynamical systems, an orbit is a collection of points related by the evolution function of the dynamical system. The orbit is a subset of the phase space and the set of all orbits is a partition of the phase space, that is different orbits do not intersect in the phase space. Understanding the properties of orbits by using topological method is one of the objectives of the modern theory of dynamical systems.
Circular orbit	A circular orbit is the orbit of any point of an object rotating around a fixed axis.
	Below we consider a circular orbit in astrodynamics or celestial mechanics under standard assumptions. Here the centripetal force is the gravitational force, and the axis mentioned above is the line through the center of the central mass perpendicular to the plane of motion.
Orbit	In mathematics, in the study of dynamical systems, an orbit is a collection of points related by the evolution function of the dynamical system. The orbit is a subset of the phase space and the set of all orbits is a partition of the phase space, that is different orbits do not intersect in the phase space. Understanding the properties of orbits by using topological method is one of the objectives of the modern theory of dynamical systems.

Chapter 7. RELATIVE MOTION AND RENDEZVOUS

Chapter 8. INTERPLANETARY TRAJECTORIES

Bi-elliptic transfer	In astronautics and aerospace engineering, the bi-elliptic transfer is an orbital maneuver that moves a spacecraft from one orbit to another and may, in certain situations, require less delta-v than a Hohmann transfer.
	The bi-elliptic transfer consists of two half elliptic orbits. From the initial orbit, a delta-v is applied boosting the spacecraft into the first transfer orbit with an apoapsis at some point r_b away from the central body.
Ellipse	In geometry, an ellipse is a plane curve that results from the intersection of a cone by a plane in a way that produces a closed curve. Circles are special cases of ellipses, obtained when the cutting plane is orthogonal to the cone's axis. An ellipse is also the locus of all points of the plane whose distances to two fixed points add to the same constant.
Circular orbit	A circular orbit is the orbit of any point of an object rotating around a fixed axis.
	Below we consider a circular orbit in astrodynamics or celestial mechanics under standard assumptions. Here the centripetal force is the gravitational force, and the axis mentioned above is the line through the center of the central mass perpendicular to the plane of motion.
Orbit	In mathematics, in the study of dynamical systems, an orbit is a collection of points related by the evolution function of the dynamical system. The orbit is a subset of the phase space and the set of all orbits is a partition of the phase space, that is different orbits do not intersect in the phase space. Understanding the properties of orbits by using topological method is one of the objectives of the modern theory of dynamical systems.
Orbit	In mathematics, in the study of dynamical systems, an orbit is a collection of points related by the evolution function of the dynamical system. The orbit is a subset of the phase space and the set of all orbits is a partition of the phase space, that is different orbits do not intersect in the phase space. Understanding the properties of orbits by using topological method is one of the objectives of the modern theory of dynamical systems.

Chapter 8. INTERPLANETARY TRAJECTORIES

Chapter 8. INTERPLANETARY TRAJECTORIES

Sphere	A sphere is a perfectly round geometrical object in three-dimensional space, such as the shape of a round ball. Like a circle in two dimensions, a perfect sphere is completely symmetrical around its center, with all points on the surface lying the same distance r from the center point. This distance r is known as the radius of the sphere.
Sphere of influence	A sphere of influence in astrodynamics and astronomy is the spherical region around a celestial body where the primary gravitational influence on an orbiting object is that body. This is usually used to describe the areas in our solar system where planets dominate the orbits of surrounding objects (such as moons), despite the presence of the much more massive (but distant) Sun. In a more general sense, the patched conic approximation is only valid within the Sphere of influence.
	The general equation describing the radius of the sphere $r_{\text{Sphere of influence}}$ of a planet:
	$$r_{SOI} = a_p \left(\frac{m_p}{m_s}\right)^{2/5}$$
	where
	In the patched conic approximation, once an object leaves the planet's Sphere of influence, the primary/only gravitational influence is the Sun (until the object enters another body's Sphere of influence).
Force	In physics, a force is any influence that causes a free body to undergo a change in speed, a change in direction, or a change in shape. Force can also be described by intuitive concepts such as a push or pull that can cause an object with mass to change its velocity (which includes to begin moving from a state of rest), i.e., to accelerate, or which can cause a flexible object to deform. A force has both magnitude and direction, making it a vector quantity.
Gravitation	In physics, Gravitation is a very important reference book on Einstein's theory of gravity by Charles W. Misner, Kip S. Thorne, and John Archibald Wheeler. Often considered the "Bible" of General Relativity by researchers for its prominence, it is frequently called MTW after its authors' initials, or "the Phone Book" due to its immense size. It was originally published by W. H. Freeman and Company in 1973.

Chapter 8. INTERPLANETARY TRAJECTORIES

Chapter 8. INTERPLANETARY TRAJECTORIES

Motion	In physics, motion is a change in position of an object with respect to time. Change in action is the result of an unapplied force. Motion is typically described in terms of velocity, acceleration, displacement, and time.
Perturbation	Perturbation is a term used in astronomy in connection with descriptions of the complex motion of a massive body which is subject to appreciable gravitational effects from more than one other massive body. Such complex motions of a body can be broken down descriptively into component parts. First, there can be the hypothetical motion that the body would follow, if it moved under the gravitational effect of one other body only.
Hypothetico-deductive model	The hypothetico-deductive model, first so-named by William Whewell, is a proposed description of scientific method. According to it, scientific inquiry proceeds by formulating a hypothesis in a form that could conceivably be falsified by a test on observable data. A test that could and does run contrary to predictions of the hypothesis is taken as a falsification of the hypothesis.
Hyperbola	In mathematics a hyperbola is a curve, specifically a smooth curve that lies in a plane, which can be defined either by its geometric properties or by the kinds of equations for which it is the solution set. A hyperbola has two pieces, called connected components or branches, which are mirror images of each other and resembling two infinite bows. The hyperbola is one of the four kinds of conic section, formed by the intersection of a plane and a cone.
Parking orbit	A parking orbit is a temporary orbit used during the launch of a satellite or other space probe. A launch vehicle boosts into the parking orbit, then coasts for a while, then fires again to enter the final desired trajectory. The alternative to a parking orbit is direct injection, where the rocket fires continuously (except during staging) until its fuel is exhausted, ending with the payload on the final trajectory.

Chapter 8. INTERPLANETARY TRAJECTORIES

Chapter 8. INTERPLANETARY TRAJECTORIES

Delta-v

In astrodynamics, Δv or delta-v is a scalar which takes units of speed that measures the amount of "effort" needed to carry out an orbital maneuver, i.e., to change from one trajectory to another.

$$\Delta v = \int_{t_0}^{t_1} \frac{|T|}{m} \, dt$$

where

T is the instantaneous thrust
m is the instantaneous mass

If there are no other external forces than gravity, this is the integral of the magnitude of the g-force.

In the absence of external forces, and when thrust is applied in a constant direction this simplifies to:

$$= \int_{t_0}^{t_1} |a| \, dt = |v_1 - v_0|$$

which is simply the magnitude of the change in velocity.

Eccentricity

In mathematics, the eccentricity, denoted e or ε, is a parameter associated with every conic section. It can be thought of as a measure of how much the conic section deviates from being circular.

Chapter 8. INTERPLANETARY TRAJECTORIES

Chapter 8. INTERPLANETARY TRAJECTORIES

In particular,

- The eccentricity of a circle is zero.
- The eccentricity of an ellipse which is not a circle is greater than zero but less than 1.
- The eccentricity of a parabola is 1.
- The eccentricity of a hyperbola is greater than 1.

Furthermore, two conic sections are similar if and only if they have the same eccentricity.

Lagrange multiplier

In mathematical optimization, the method of Lagrange multipliers provides a strategy for finding the maxima and minima of a function subject to constraints.

For instance, consider the optimization problem

$$\text{maximize } f(x,y)$$
$$\text{subject to } g(x,y) = c.$$

We introduce a new variable (λ) called a Lagrange multiplier, and study the Lagrange function defined by

$$\Lambda(x, y, \lambda) = f(x,y) + \lambda \cdot \big(g(x,y) - c\big).$$

(the λ term may be either added or subtracted). If (x,y) is a maximum for the original constrained problem, then there exists a λ such that (x,y,λ) is a stationary point for the Lagrange function (stationary points are those points where the partial derivatives of Λ are zero).

Multiplier

In Fourier analysis, a multiplier operator is a type of linear operator, or transformation of functions. These operators act on a function by altering its Fourier transform. Specifically they multiply the Fourier transform of a function by a specified function known as the multiplier or symbol.

Chapter 8. INTERPLANETARY TRAJECTORIES

Chapter 8. INTERPLANETARY TRAJECTORIES

Orbital elements

Orbital elements are the parameters required to uniquely identify a specific orbit. In celestial mechanics these elements are generally considered in classical two-body systems, where a Kepler orbit is used (derived from Newton's laws of motion and Newton's law of universal gravitation). There are many different ways to mathematically describe the same orbit, but certain schemes each consisting of a set of six parameters are commonly used in astronomy and orbital mechanics.

Flattening

The flattening, ellipticity, or oblateness of an oblate spheroid is a measure of the "squashing" of the spheroid's pole, towards its equator. If a is the distance from the spheroid center to the equator and b the distance from the center to the pole then

$$flattening = \frac{a-b}{a}$$

First, second and third flattening

The first, primary flattening, f, is the versine of the spheroid's angular eccentricity (" α "), equalling the relative difference between its equatorial radius, a, and its polar radius, b:

$$f = \text{ver}(\alpha) = 2\sin^2\left(\frac{\alpha}{2}\right) = 1 - \cos(\alpha) = \frac{a-b}{a};$$

- The flattening of the Earth in WGS-84 is 1:298.257223563 (which corresponds to a radius difference of 21.385 km (13 mi) of the Earth radius 6378.137 - 6356.752 km) and would not be realized visually from space, since the difference represents only 0.335 %.
- The flattening of Jupiter (1:16) and Saturn (nearly 1:10), in contrast, can be seen even in a small telescope;
- Conversely, that of the Sun is less than 1:1000 and that of the Moon barely 1:900.

The amount of flattening depends on

Chapter 8. INTERPLANETARY TRAJECTORIES

Chapter 8. INTERPLANETARY TRAJECTORIES

and in detail on

- size and density of the celestial body ;
- the rotation of the planet or star;
- and the elasticity of the body.

There is also a second flattening, f',

$$f' = \frac{2\sin^2(\alpha/2)}{1 - 2\sin^2(\alpha/2)} = \frac{a-b}{b}$$

and a third flattening, f'' (sometimes denoted as "n", firstly used the notation in 1837 by Friedrich Bessel on calculation of meridian arc length), that is the squared half-angle tangent of α:

$$f'' = n = \tan^2\left(\frac{\alpha}{2}\right) = \frac{1 - \cos(\alpha)}{1 + \cos(\alpha)} = \frac{a-b}{a+b};$$

Inclination

Inclination in general is the angle between a reference plane and another plane or axis of direction.

Orbits

The inclination is one of the six orbital parameters describing the shape and orientation of a celestial orbit. It is the angular distance of the orbital plane from the plane of reference (usually the primary's equator or the ecliptic), normally stated in degrees.

Ephemeris

An ephemeris is a table of values that gives the positions of astronomical objects in the sky at a given time or times. Different kinds of ephemerides are used for astronomy and astrology. Even though this was also one of the first applications of mechanical computers, an ephemeris will still often be a simple printed table.

Chapter 8. INTERPLANETARY TRAJECTORIES

Chapter 8. INTERPLANETARY TRAJECTORIES

Epoch

In astronomy, an epoch is a moment in time used as a reference point for some time-varying astronomical quantity, such as celestial coordinates, or elliptical orbital elements of a celestial body, where these are (as usual) subject to perturbations and vary with time. The time-varying astronomical quantities might include, for example, the mean longitude or mean anomaly of a body, or of the node of its orbit relative to a reference-plane, or of the direction of the apogee or aphelion of its orbit, or the size of the major axis of its orbit.

The main uses of astronomical quantities specified in this way include their use to calculate other parameters of relevant motions, e.g. in order to predict future positions and velocities.

Julian day

Julian day is used in the Julian date system of time measurement for scientific use by the astronomy community. Julian date is recommended for astronomical use by the International Astronomical Union.

Julian Date

Historical Julian dates were recorded relative to GMT or Ephemeris Time, but the International Astronomical Union now recommends that Julian Dates be specified in Terrestrial Time, and that when necessary to specify Julian Dates using a different time scale, that the time scale used be indicated when required, such as Julian day(UT1).

State

In functional analysis, a state on a C*-algebra is a positive linear functional of norm 1. The set of states of a C*-algebra A, sometimes denoted by S(A), is always a convex set. The extremal points of S(A) are called pure states. If A has a multiplicative identity, S(A) is compact in the weak*-topology.

Chapter 8. INTERPLANETARY TRAJECTORIES

Chapter 9. RIGID-BODY DYNAMICS

Kinematics

It is natural to begin this discussion by considering the various possible types of motion in themselves, leaving out of account for a time the causes to which the initiation of motion may be ascribed; this preliminary enquiry constitutes the science of Kinematics.

--ET Whittaker

Kinematics is the branch of classical mechanics that describes the motion of bodies (objects) and systems (groups of objects) without consideration of the forces that cause the motion.

Kinematics is not to be confused with another branch of classical mechanics: analytical dynamics (the study of the relationship between the motion of objects and its causes), sometimes subdivided into kinetics (the study of the relation between external forces and motion) and statics (the study of the relations in a system at equilibrium). Kinematics also differs from dynamics as used in modern-day physics to describe time-evolution of a system.

Relative velocity

In non-relativistic kinematics, relative velocity is the vector difference between the velocities of two objects, as evaluated in terms of a single coordinate system.

For example, if the velocities of particles A and B are $\mathbf{V}_A$ and $\mathbf{V}_B$ respectively in terms of a given coordinate system, then the relative velocity of A with respect to B (also called the velocity of A relative to B, $\mathbf{V}_{A/B}$, or $\mathbf{V}_{A\ rel\ B}$) is

$$\mathbf{V}_{A\ rel\ B} = \mathbf{V}_A - \mathbf{V}_B.$$

Conversely, the velocity of B relative to A is

$$\mathbf{V}_{B\ rel\ A} = \mathbf{V}_B - \mathbf{V}_A.$$

The expression "the velocity of A relative to B" is shorthand for "the velocity of A in the coordinate system where B is always at rest".

Chapter 9. RIGID-BODY DYNAMICS

Chapter 9. RIGID-BODY DYNAMICS

Circular orbit	A circular orbit is the orbit of any point of an object rotating around a fixed axis.
	Below we consider a circular orbit in astrodynamics or celestial mechanics under standard assumptions. Here the centripetal force is the gravitational force, and the axis mentioned above is the line through the center of the central mass perpendicular to the plane of motion.
Orbit	In mathematics, in the study of dynamical systems, an orbit is a collection of points related by the evolution function of the dynamical system. The orbit is a subset of the phase space and the set of all orbits is a partition of the phase space, that is different orbits do not intersect in the phase space. Understanding the properties of orbits by using topological method is one of the objectives of the modern theory of dynamical systems.
Inertia	Inertia is the resistance of any physical object to a change in its state of motion or rest. It is represented numerically by an object's mass. The principle of inertia is one of the fundamental principles of classical physics which are used to describe the motion of matter and how it is affected by applied forces.
Angular velocity	In physics, the angular velocity is a vector quantity (more precisely, a pseudovector) which specifies the angular speed of an object and the axis about which the object is rotating. The SI unit of angular velocity is radians per second, although it may be measured in other units such as degrees per second, revolutions per second, degrees per hour, etc. When measured in cycles or rotations per unit time (e.g. revolutions per minute), it is often called the rotational velocity and its magnitude the rotational speed.
Momentum	In classical mechanics, Momentum is the product of the mass and velocity of an object . In relativistic mechanics, this quantity is multiplied by the Lorentz factor. Momentum is sometimes referred to as linear Momentum to distinguish it from the related subject of angular Momentum.
Continuous	In probability theory, a probability distribution is called continuous if its cumulative distribution function is continuous. This is equivalent to saying that for random variables X with the distribution in question, Pr[X = a] = 0 for all real numbers a, i.e.: the probability that X attains the value a is zero, for any number a. If the distribution of X is continuous then X is called a continuous random variable.

Chapter 9. RIGID-BODY DYNAMICS

Chapter 9. RIGID-BODY DYNAMICS

Motion	In physics, motion is a change in position of an object with respect to time. Change in action is the result of an unapplied force. Motion is typically described in terms of velocity, acceleration, displacement, and time.
Fluid	In physics, a fluid is a substance that continually deforms (flows) under an applied shear stress, no matter how small. Fluids are a subset of the phases of matter and include liquids, gases, plasmas and, to some extent, plastic solids.
	In common usage, "fluid" is often used as a synonym for "liquid", with no implication that gas could also be present.
Angular momentum	In physics, angular momentum, moment of momentum, or rotational momentum is a conserved vector quantity that can be used to describe the overall state of a physical system. The angular momentum L of a particle with respect to some point of origin is $$\mathbf{L} = \mathbf{r} \times \mathbf{p} = \mathbf{r} \times m\mathbf{v},$$ where r is the particle's position from the origin, p = mv is its linear momentum, and × denotes the cross product.
	The angular momentum of a system of particles (e.g. a rigid body) is the sum of angular momenta of the individual particles.
Impulse	In classical mechanics, an impulse is defined as the integral of a force with respect to time. When a force is applied to a rigid body it changes the momentum of that body. A small force applied for a long time can produce the same momentum change as a large force applied briefly, because it is the product of the force and the time for which it is applied that is important.
Parallel axis theorem	In physics, the parallel axis theorem, given the moment of inertia of the object about the parallel axis through the object's centre of mass and the perpendicular distance (r) between the axes.

Chapter 9. RIGID-BODY DYNAMICS

Chapter 9. RIGID-BODY DYNAMICS

The moment of inertia about the new axis z is given by:

$$I_z = I_{cm} + mr^2,$$

where:

I_{cm} is the moment of inertia of the object about an axis passing through its centre of mass;
m is the object's mass;
r is the perpendicular distance between the two axes.

This rule can be applied with the stretch rule and perpendicular axis theorem to find moments of inertia for a variety of shapes.

The parallel axes rule also applies to the second moment of area (area moment of inertia) for a plane region D:

$$I_z = I_x + Ar^2,$$

where:

I_z is the area moment of inertia of D relative to the parallel axis;
I_x is the area moment of inertia of D relative to its centroid;
A is the area of the plane region D;
r is the distance from the new axis z to the centroid of the plane region D.

Chapter 9. RIGID-BODY DYNAMICS

Chapter 9. RIGID-BODY DYNAMICS

Note: The centroid of D coincides with the centre of gravity (CG) of a physical plate with the same shape that has uniform density.

Orbit

In mathematics, in the study of dynamical systems, an orbit is a collection of points related by the evolution function of the dynamical system. The orbit is a subset of the phase space and the set of all orbits is a partition of the phase space, that is different orbits do not intersect in the phase space. Understanding the properties of orbits by using topological method is one of the objectives of the modern theory of dynamical systems.

Precession

Precession is the process of a round part in a round hole rotating with respect to that hole because of clearance between between them and a radial force on the part than changes direction. The direction of rotation of the inner part is opposite to the direction of rotation of the radial force. Fretting between the part and the hole is often a result of this motion.

Euler angles

The Euler angles are three angles introduced by Leonhard Euler to describe the orientation of a rigid body. To describe such an orientation in 3-dimensional Euclidean space three parameters are required. They can be given in several ways, Euler angles being one of them; see charts on SO(3) for others.

Delta-v

In astrodynamics, Δv or delta-v is a scalar which takes units of speed that measures the amount of "effort" needed to carry out an orbital maneuver, i.e., to change from one trajectory to another.

$$\Delta v = \int_{t_0}^{t_1} \frac{|T|}{m} dt$$

where

T is the instantaneous thrust
m is the instantaneous mass

If there are no other external forces than gravity, this is the integral of the magnitude of the g-force.

Chapter 9. RIGID-BODY DYNAMICS

Chapter 9. RIGID-BODY DYNAMICS

In the absence of external forces, and when thrust is applied in a constant direction this simplifies to:

$$= \int_{t_0}^{t_1} |a|\, dt = |v_1 - v_0|$$

which is simply the magnitude of the change in velocity.

Rotation — In geometry and linear algebra, a rotation is a transformation in a plane or in space that describes the motion of a rigid body around a fixed point. A rotation is different from a translation, which has no fixed points, and from a reflection, which "flips" the bodies it is transforming. A rotation and the above-mentioned transformations are isometries; they leave the distance between any two points unchanged after the transformation.

Force — In physics, a force is any influence that causes a free body to undergo a change in speed, a change in direction, or a change in shape. Force can also be described by intuitive concepts such as a push or pull that can cause an object with mass to change its velocity (which includes to begin moving from a state of rest), i.e., to accelerate, or which can cause a flexible object to deform. A force has both magnitude and direction, making it a vector quantity.

Chapter 9. RIGID-BODY DYNAMICS

Chapter 10. SATELLITE ATTITUDE DYNAMICS

Stability	In probability theory, the stability of a random variable is the property that a linear combination of two independent copies of the variable has the same distribution, up to location and scale parameters. The distributions of random variables having this property are said to be "stable distributions". Results available in probability theory show that all possible distributions having this property are members of a four-parameter family of distributions.
Orbit	In mathematics, in the study of dynamical systems, an orbit is a collection of points related by the evolution function of the dynamical system. The orbit is a subset of the phase space and the set of all orbits is a partition of the phase space, that is different orbits do not intersect in the phase space. Understanding the properties of orbits by using topological method is one of the objectives of the modern theory of dynamical systems.
Orbit determination	Orbit determination is a branch of astronomy specialised in calculating, and hence predicting, the orbits of objects such as moons, planets, and spacecraft . These orbits could be orbiting the Earth, or other bodies. The determination of the orbit of newly observed asteroids is a common usage of these techniques, both so the asteroid can be followed up with future observations, and also to check that it has not been previously discovered.
Spinner	A spinner is a method of randomly determining something in a game, often movement. There are two types of spinners: 1. A balanced arrow anchored to a hard backing, usually cardboard, with printed sections. A spinner is used in some games instead of Dice because a spinner provides more space to print text or images.
Motion	In physics, motion is a change in position of an object with respect to time. Change in action is the result of an unapplied force. Motion is typically described in terms of velocity, acceleration, displacement, and time.
Angular velocity	In physics, the angular velocity is a vector quantity (more precisely, a pseudovector) which specifies the angular speed of an object and the axis about which the object is rotating. The SI unit of angular velocity is radians per second, although it may be measured in other units such as degrees per second, revolutions per second, degrees per hour, etc. When measured in cycles or rotations per unit time (e.g. revolutions per minute), it is often called the rotational velocity and its magnitude the rotational speed.

Chapter 10. SATELLITE ATTITUDE DYNAMICS

Chapter 10. SATELLITE ATTITUDE DYNAMICS

Precession	Precession is the process of a round part in a round hole rotating with respect to that hole because of clearance between between them and a radial force on the part than changes direction. The direction of rotation of the inner part is opposite to the direction of rotation of the radial force. Fretting between the part and the hole is often a result of this motion.
Orbit	In mathematics, in the study of dynamical systems, an orbit is a collection of points related by the evolution function of the dynamical system. The orbit is a subset of the phase space and the set of all orbits is a partition of the phase space, that is different orbits do not intersect in the phase space. Understanding the properties of orbits by using topological method is one of the objectives of the modern theory of dynamical systems.
Perturbation	Perturbation is a term used in astronomy in connection with descriptions of the complex motion of a massive body which is subject to appreciable gravitational effects from more than one other massive body.
	Such complex motions of a body can be broken down descriptively into component parts. First, there can be the hypothetical motion that the body would follow, if it moved under the gravitational effect of one other body only.
Gravitation	In physics, Gravitation is a very important reference book on Einstein's theory of gravity by Charles W. Misner, Kip S. Thorne, and John Archibald Wheeler. Often considered the "Bible" of General Relativity by researchers for its prominence, it is frequently called MTW after its authors' initials, or "the Phone Book" due to its immense size. It was originally published by W. H. Freeman and Company in 1973.
Inertia	Inertia is the resistance of any physical object to a change in its state of motion or rest. It is represented numerically by an object's mass. The principle of inertia is one of the fundamental principles of classical physics which are used to describe the motion of matter and how it is affected by applied forces.
Momentum	In classical mechanics, Momentum is the product of the mass and velocity of an object . In relativistic mechanics, this quantity is multiplied by the Lorentz factor. Momentum is sometimes referred to as linear Momentum to distinguish it from the related subject of angular Momentum.
Circular orbit	A circular orbit is the orbit of any point of an object rotating around a fixed axis.

Chapter 10. SATELLITE ATTITUDE DYNAMICS

Chapter 10. SATELLITE ATTITUDE DYNAMICS

	Below we consider a circular orbit in astrodynamics or celestial mechanics under standard assumptions. Here the centripetal force is the gravitational force, and the axis mentioned above is the line through the center of the central mass perpendicular to the plane of motion.
Dissipation	In physics, dissipation embodies the concept of a dynamical system where important mechanical models, such as waves or oscillations, lose energy over time, typically from friction or turbulence. The lost energy converts into heat, which raises the temperature of the system. Such systems are called dissipative systems.
Force	In physics, a force is any influence that causes a free body to undergo a change in speed, a change in direction, or a change in shape. Force can also be described by intuitive concepts such as a push or pull that can cause an object with mass to change its velocity (which includes to begin moving from a state of rest), i.e., to accelerate, or which can cause a flexible object to deform. A force has both magnitude and direction, making it a vector quantity.
Spinner	A spinner is a method of randomly determining something in a game, often movement. There are two types of spinners: 1. A balanced arrow anchored to a hard backing, usually cardboard, with printed sections. A spinner is used in some games instead of Dice because a spinner provides more space to print text or images.
Impulse	In classical mechanics, an impulse is defined as the integral of a force with respect to time. When a force is applied to a rigid body it changes the momentum of that body. A small force applied for a long time can produce the same momentum change as a large force applied briefly, because it is the product of the force and the time for which it is applied that is important.
Spin	In quantum mechanics and particle physics, spin is a fundamental characteristic property of elementary particles, composite particles (hadrons), and atomic nuclei.

Chapter 10. SATELLITE ATTITUDE DYNAMICS

Chapter 10. SATELLITE ATTITUDE DYNAMICS

All elementary particles of a given kind have the same spin quantum number, an important part of a particle's quantum state. When combined with the spin-statistics theorem, the spin of electrons results in the Pauli exclusion principle, which in turn underlies the periodic table of chemical elements.

Angular momentum

In physics, angular momentum, moment of momentum, or rotational momentum is a conserved vector quantity that can be used to describe the overall state of a physical system. The angular momentum L of a particle with respect to some point of origin is

$$\mathbf{L} = \mathbf{r} \times \mathbf{p} = \mathbf{r} \times m\mathbf{v},$$

where r is the particle's position from the origin, p = mv is its linear momentum, and × denotes the cross product.

The angular momentum of a system of particles (e.g. a rigid body) is the sum of angular momenta of the individual particles.

Mechanism

Mechanism is the belief that natural wholes (principally living things) are like machines or artifacts, composed of parts lacking any intrinsic relationship to each other, and with their order imposed from without. Thus, the source of an apparent thing's activities is not the whole itself, but its parts or an external influence on the parts. Mechanism is opposed to the organic conception of nature best articulated by Aristotle and more recently elaborated as vitalism.

Inclination

Inclination in general is the angle between a reference plane and another plane or axis of direction.

Orbits

The inclination is one of the six orbital parameters describing the shape and orientation of a celestial orbit. It is the angular distance of the orbital plane from the plane of reference (usually the primary's equator or the ecliptic), normally stated in degrees.

Chapter 10. SATELLITE ATTITUDE DYNAMICS

Chapter 10. SATELLITE ATTITUDE DYNAMICS

Gravity assist

In orbital mechanics and aerospace engineering, a gravitational slingshot, gravity assist maneuver, or swing-by is the use of the relative movement and gravity of a planet or other celestial body to alter the path and speed of a spacecraft, typically in order to save propellant, time, and expense. Gravity assistance can be used to accelerate, decelerate and/or re-direct the path of a spacecraft.

The "assist" is provided by the motion (orbital angular momentum) of the gravitating body as it pulls on the spacecraft.

Hypothetico-deductive model

The hypothetico-deductive model, first so-named by William Whewell, is a proposed description of scientific method. According to it, scientific inquiry proceeds by formulating a hypothesis in a form that could conceivably be falsified by a test on observable data. A test that could and does run contrary to predictions of the hypothesis is taken as a falsification of the hypothesis.

Frequency

Frequency is the number of occurrences of a repeating event per unit time. It is also referred to as temporal frequency. The period is the duration of one cycle in a repeating event, so the period is the reciprocal of the frequency.

Oscillation

Oscillation is the repetitive variation, typically in time, of some measure about a central value (often a point of equilibrium) or between two or more different states. Familiar examples include a swinging pendulum and AC power. The term vibration is sometimes used more narrowly to mean a mechanical oscillation but sometimes is used to be synonymous with "oscillation".

Delta-v

In astrodynamics, Δv or delta-v is a scalar which takes units of speed that measures the amount of "effort" needed to carry out an orbital maneuver, i.e., to change from one trajectory to another.

$$\Delta v = \int_{t_0}^{t_1} \frac{|T|}{m} \, dt$$

Chapter 10. SATELLITE ATTITUDE DYNAMICS

where

> T is the instantaneous thrust
> m is the instantaneous mass

If there are no other external forces than gravity, this is the integral of the magnitude of the g-force.

In the absence of external forces, and when thrust is applied in a constant direction this simplifies to:

$$= \int_{t_0}^{t_1} |a|\, dt = |v_1 - v_0|$$

which is simply the magnitude of the change in velocity.

Chapter 10. SATELLITE ATTITUDE DYNAMICS

Chapter 11. ROCKET VEHICLE DYNAMICS

Gravity turn	A gravity turn is a maneuver used in launching a spacecraft into, or descending from, an orbit around a celestial body such as a planet or a moon. This launch trajectory offers two main advantages over a thrust-controlled trajectory where the rocket's own thrust steers the vehicle. First, any thrust used to change the ship's direction does not accelerate the vehicle into orbit.
Impulse	In classical mechanics, an impulse is defined as the integral of a force with respect to time. When a force is applied to a rigid body it changes the momentum of that body. A small force applied for a long time can produce the same momentum change as a large force applied briefly, because it is the product of the force and the time for which it is applied that is important.
Specific impulse	Specific impulse is a way to describe the efficiency of rocket and jet engines. It represents the impulse (change in momentum) per unit amount of propellant used. The unit amount may be given either per unit mass (such as kilograms), or per unit Earth-weight (such as kiloponds, since g is used for the latter definition).
Equations of motion	Equations of motion are equations that describe the behavior of a system (e.g., the motion of a particle under the influence of a force) as a function of time. Sometimes the term refers to the differential equations that the system satisfies (e.g., Newton's second law or Euler-Lagrange equations), and sometimes to the solutions to those equations. Equations of uniformly accelerated linear motion The equations that apply to bodies moving linearly (in one dimension) with constant acceleration are often referred to as "SUVAT" equations where the five variables are represented by those letters (s = displacement, u = initial velocity, v = final velocity, a = acceleration, t = time); the five letters may be shown in a different order.
Gravitation	In physics, Gravitation is a very important reference book on Einstein's theory of gravity by Charles W. Misner, Kip S. Thorne, and John Archibald Wheeler. Often considered the "Bible" of General Relativity by researchers for its prominence, it is frequently called MTW after its authors' initials, or "the Phone Book" due to its immense size. It was originally published by W. H. Freeman and Company in 1973.
Motion	In physics, motion is a change in position of an object with respect to time. Change in action is the result of an unapplied force. Motion is typically described in terms of velocity, acceleration, displacement, and time.

Chapter 11. ROCKET VEHICLE DYNAMICS

Chapter 11. ROCKET VEHICLE DYNAMICS

Orbital maneuver	In spaceflight, an orbital maneuver is the use of propulsion systems to change the orbit of a spacecraft. For spacecraft far from Earth--for example those in orbits around the Sun--an orbital maneuver is called a deep-space maneuver (DSM). Impulsive maneuvers An "impulsive maneuver" is one which involves a single, ideally instantaneous change in the spacecraft's velocity.
Curvature	In mathematics, curvature refers to any of a number of loosely related concepts in different areas of geometry. Intuitively, curvature is the amount by which a geometric object deviates from being flat, or straight in the case of a line, but this is defined in different ways depending on the context. There is a key distinction between extrinsic curvature, which is defined for objects embedded in another space (usually an Euclidean space) in a way that relates to the radius of curvature of circles that touch the object, and intrinsic curvature, which is defined at each point in a Riemannian manifold.
Force	In physics, a force is any influence that causes a free body to undergo a change in speed, a change in direction, or a change in shape. Force can also be described by intuitive concepts such as a push or pull that can cause an object with mass to change its velocity (which includes to begin moving from a state of rest), i.e., to accelerate, or which can cause a flexible object to deform. A force has both magnitude and direction, making it a vector quantity.
Altitude	Altitude is used (aviation, geometry, geographical survey, sport, and more). As a general definition, altitude is a distance measurement, usually in the vertical or "up" direction, between a reference datum and a point or object. The reference datum also often varies according to the context.
Lagrange multiplier	In mathematical optimization, the method of Lagrange multipliers provides a strategy for finding the maxima and minima of a function subject to constraints.

Chapter 11. ROCKET VEHICLE DYNAMICS

Chapter 11. ROCKET VEHICLE DYNAMICS

For instance, consider the optimization problem

$$\begin{aligned} \text{maximize } & f(x,y) \\ \text{subject to } & g(x,y) = c. \end{aligned}$$

We introduce a new variable (λ) called a Lagrange multiplier, and study the Lagrange function defined by

$$\Lambda(x,y,\lambda) = f(x,y) + \lambda \cdot \Big(g(x,y) - c\Big).$$

(the λ term may be either added or subtracted). If (x,y) is a maximum for the original constrained problem, then there exists a λ such that (x,y,λ) is a stationary point for the Lagrange function (stationary points are those points where the partial derivatives of Λ are zero).

Momentum	In classical mechanics, Momentum is the product of the mass and velocity of an object. In relativistic mechanics, this quantity is multiplied by the Lorentz factor. Momentum is sometimes referred to as linear Momentum to distinguish it from the related subject of angular Momentum.
Hypothetico-deductive model	The hypothetico-deductive model, first so-named by William Whewell, is a proposed description of scientific method. According to it, scientific inquiry proceeds by formulating a hypothesis in a form that could conceivably be falsified by a test on observable data. A test that could and does run contrary to predictions of the hypothesis is taken as a falsification of the hypothesis.
Multiplier	In Fourier analysis, a multiplier operator is a type of linear operator, or transformation of functions. These operators act on a function by altering its Fourier transform. Specifically they multiply the Fourier transform of a function by a specified function known as the multiplier or symbol.
Series	A series is the sum of the terms of a sequence. Finite sequences and series have defined first and last terms, whereas infinite sequences and series continue indefinitely.

Chapter 11. ROCKET VEHICLE DYNAMICS

Chapter 11. ROCKET VEHICLE DYNAMICS

In mathematics, given an infinite sequence of numbers { a_n }, a series is informally the result of adding all those terms together: $a_1 + a_2 + a_3 + \cdots$.

Tandem

Tandem is an arrangement where a team of machines, animals or people are lined up one behind another, all facing in the same direction.

Tandem harness is used for two or more draft horses (or other draft animals) harnessed in a single line one behind another, as opposed to a pair, harnessed side-by-side, or a team of several pairs. Tandem harness allows additional animals to provide pulling power for a vehicle designed for a single animal.

Numerical integration

In numerical analysis, numerical integration constitutes a broad family of algorithms for calculating the numerical value of a definite integral, and by extension, the term is also sometimes used to describe the numerical solution of differential equations The term numerical quadrature is more or less a synonym for numerical integration, especially as applied to one-dimensional integrals.

Stumpff function

In celestial mechanics, the Stumpff functions $c_k(x)$, developed by Karl Stumpff, are used for analyzing orbits using the universal variable formulation. They are defined by the formula:

$$c_k(x) = \frac{1}{k!} - \frac{x}{(k+2)!} + \frac{x^2}{(k+4)!} - \ldots = \sum_{i=0}^{\infty} \frac{(-1)^i x^i}{(k+2i)!}$$

for k = 0,1,2,3.... The series above converges absolutely for all real x.

By comparing the Taylor series expansion of the trigonometric functions sin and cos with $c_0(x)$ and $c_1(x)$, a relationship can be found:

Similarly, by comparing with the expansion of the hyperbolic functions sinh and cosh we find:

Chapter 11. ROCKET VEHICLE DYNAMICS

Chapter 11. ROCKET VEHICLE DYNAMICS

	The Stumpff functions satisfy the recursive relations:
Eccentric anomaly	In celestial mechanics, the eccentric anomaly is an angular parameter that defines the position of a body that is moving along an elliptic Kepler orbit. For the point P orbiting around an ellipse, the eccentric anomaly is the angle E in the figure. It is determined by drawing a vertical line from the major axis of the ellipse through the point P and locating its intercept P′ with the auxiliary circle, a circle of radius a (the semi-major axis of the ellipse) that enscribes the entire ellipse.
Hyperbola	In mathematics a hyperbola is a curve, specifically a smooth curve that lies in a plane, which can be defined either by its geometric properties or by the kinds of equations for which it is the solution set. A hyperbola has two pieces, called connected components or branches, which are mirror images of each other and resembling two infinite bows. The hyperbola is one of the four kinds of conic section, formed by the intersection of a plane and a cone.
Orbit	In mathematics, in the study of dynamical systems, an orbit is a collection of points related by the evolution function of the dynamical system. The orbit is a subset of the phase space and the set of all orbits is a partition of the phase space, that is different orbits do not intersect in the phase space. Understanding the properties of orbits by using topological method is one of the objectives of the modern theory of dynamical systems.
Orbit determination	Orbit determination is a branch of astronomy specialised in calculating, and hence predicting, the orbits of objects such as moons, planets, and spacecraft . These orbits could be orbiting the Earth, or other bodies. The determination of the orbit of newly observed asteroids is a common usage of these techniques, both so the asteroid can be followed up with future observations, and also to check that it has not been previously discovered.
Derivative	In calculus, a branch of mathematics, the derivative is a measure of how a function changes as its input changes. Loosely speaking, a derivative can be thought of as how much one quantity is changing in response to changes in some other quantity; for example, the derivative of the position of a moving object with respect to time is the object's instantaneous velocity. Conversely, the integral of the object's velocity over time is how much the object's position changes from the time when the integral begins to the time when the integral ends.

Chapter 11. ROCKET VEHICLE DYNAMICS

Chapter 11. ROCKET VEHICLE DYNAMICS

State	In functional analysis, a state on a C*-algebra is a positive linear functional of norm 1. The set of states of a C*-algebra A, sometimes denoted by S(A), is always a convex set. The extremal points of S(A) are called pure states. If A has a multiplicative identity, S(A) is compact in the weak*-topology.
Orbital elements	Orbital elements are the parameters required to uniquely identify a specific orbit. In celestial mechanics these elements are generally considered in classical two-body systems, where a Kepler orbit is used (derived from Newton's laws of motion and Newton's law of universal gravitation). There are many different ways to mathematically describe the same orbit, but certain schemes each consisting of a set of six parameters are commonly used in astronomy and orbital mechanics.
Julian day	Julian day is used in the Julian date system of time measurement for scientific use by the astronomy community. Julian date is recommended for astronomical use by the International Astronomical Union. Julian Date Historical Julian dates were recorded relative to GMT or Ephemeris Time, but the International Astronomical Union now recommends that Julian Dates be specified in Terrestrial Time, and that when necessary to specify Julian Dates using a different time scale, that the time scale used be indicated when required, such as Julian day(UT1).
Epoch	In astronomy, an epoch is a moment in time used as a reference point for some time-varying astronomical quantity, such as celestial coordinates, or elliptical orbital elements of a celestial body, where these are (as usual) subject to perturbations and vary with time. The time-varying astronomical quantities might include, for example, the mean longitude or mean anomaly of a body, or of the node of its orbit relative to a reference-plane, or of the direction of the apogee or aphelion of its orbit, or the size of the major axis of its orbit. The main uses of astronomical quantities specified in this way include their use to calculate other parameters of relevant motions, e.g. in order to predict future positions and velocities.

Chapter 11. ROCKET VEHICLE DYNAMICS

Chapter 11. ROCKET VEHICLE DYNAMICS

Sphere | A sphere is a perfectly round geometrical object in three-dimensional space, such as the shape of a round ball. Like a circle in two dimensions, a perfect sphere is completely symmetrical around its center, with all points on the surface lying the same distance r from the center point. This distance r is known as the radius of the sphere.

Sphere of influence | A sphere of influence in astrodynamics and astronomy is the spherical region around a celestial body where the primary gravitational influence on an orbiting object is that body. This is usually used to describe the areas in our solar system where planets dominate the orbits of surrounding objects (such as moons), despite the presence of the much more massive (but distant) Sun. In a more general sense, the patched conic approximation is only valid within the Sphere of influence.

The general equation describing the radius of the sphere $r_{\text{Sphere of influence}}$ of a planet:

$$r_{SOI} = a_p \left(\frac{m_p}{m_s}\right)^{2/5}$$

where

In the patched conic approximation, once an object leaves the planet's Sphere of influence, the primary/only gravitational influence is the Sun (until the object enters another body's Sphere of influence).

Sphere | A sphere is a perfectly round geometrical object in three-dimensional space, such as the shape of a round ball. Like a circle in two dimensions, a perfect sphere is completely symmetrical around its center, with all points on the surface lying the same distance r from the center point. This distance r is known as the radius of the sphere.

Distribution | In differential geometry, a discipline within mathematics, a distribution is a subset of the tangent bundle of a manifold satisfying certain properties.

Chapter 11. ROCKET VEHICLE DYNAMICS

Chapter 11. ROCKET VEHICLE DYNAMICS

Hyperbola	In mathematics a hyperbola is a curve, specifically a smooth curve that lies in a plane, which can be defined either by its geometric properties or by the kinds of equations for which it is the solution set. A hyperbola has two pieces, called connected components or branches, which are mirror images of each other and resembling two infinite bows. The hyperbola is one of the four kinds of conic section, formed by the intersection of a plane and a cone.
Rotation	In geometry and linear algebra, a rotation is a transformation in a plane or in space that describes the motion of a rigid body around a fixed point. A rotation is different from a translation, which has no fixed points, and from a reflection, which "flips" the bodies it is transforming. A rotation and the above-mentioned transformations are isometries; they leave the distance between any two points unchanged after the transformation.

Chapter 11. ROCKET VEHICLE DYNAMICS